TINY CACTUS PUBLISHING

COPYRIGHT 2017 ADULT COLORING BOOK BY TINY CACTUS PUBLISHING.

ALL RIGHT RESERVED.

NO PART OF THIS BOOK MAY BE REPRODUCED TRANSMITTED, OR STORED IN ANY FORM OR BY ANY MEANS EXCEPT FOR YOUR OWN PERSONAL USE OR FOR A BOOK REVIEW, WITHOUT THE EXPRESS WRITTEN PERMISSION OF THE AUTHOR:

TINY CACTUS PUBLISHING

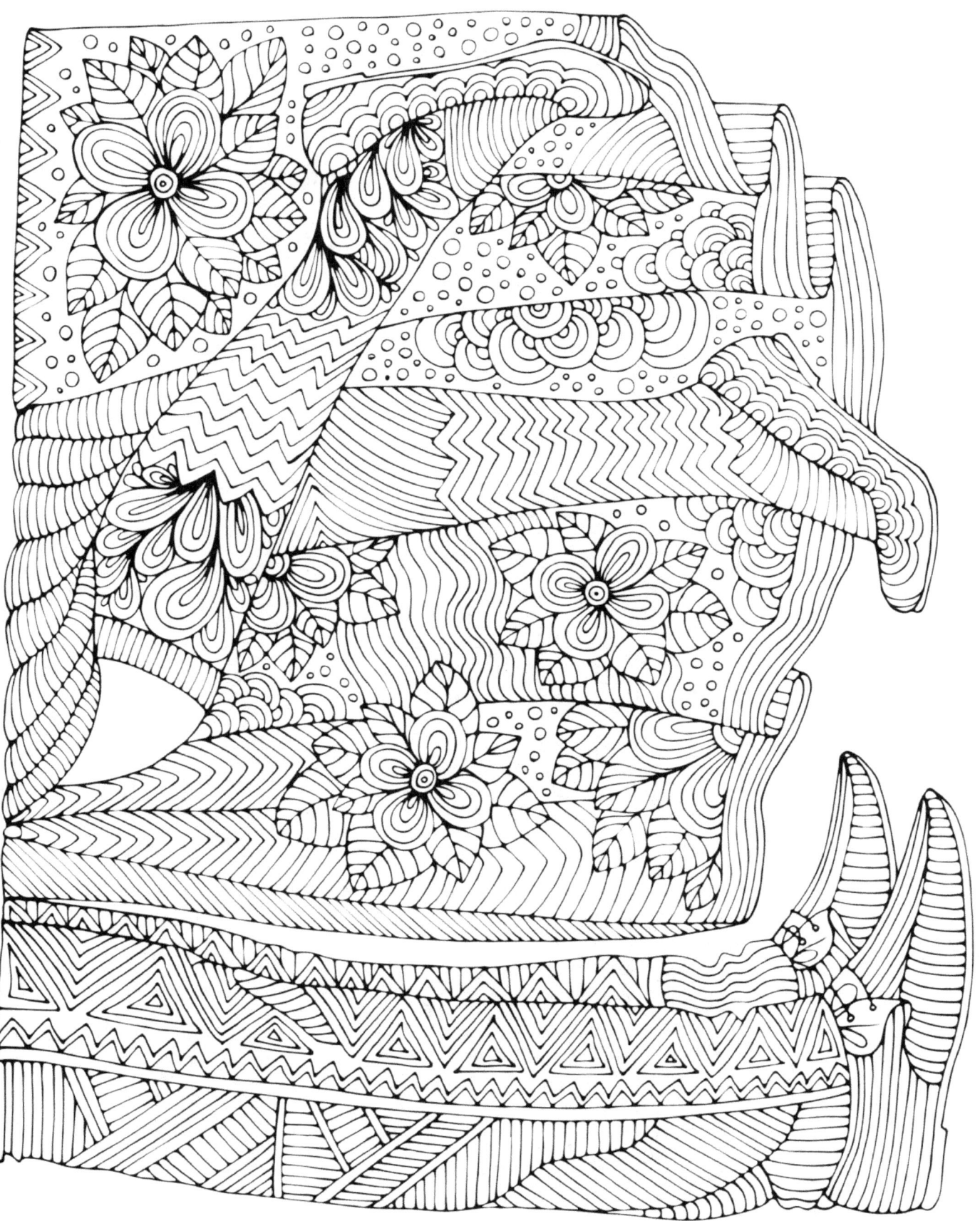